The Launch of the H3 Rocket

I0427061

Japan's Trailblazing Expedition into Space Exploration

Sam Holly

Copyright

Table of Content

Overview

Humanity has always gazed skyward at the vastness of the cosmos, filled with wonder and curiosity as it seeks to unravel its secrets. Amid the many stars and galaxies, Japan is the country leading the way in space research.

Japan has a long history of pioneering scientific advancements and a never-ending spirit of exploration, which have allowed it to continuously push the limits of what is feasible for space flight.

The H3 Launch Vehicle is the centrepiece of Japan's space programme and a colossal example of human creativity and perseverance.

The H3, which was created in collaboration between Mitsubishi Heavy Industries (MHI) and the Japan Aerospace Exploration Agency (JAXA), is the epitome of Japanese rocketry as it combines state-of-the-art technology with decades of spaceflight expertise.

The H3 is a medium-lift launch vehicle with an astounding height of 63 metres and a diameter of 5.27 metres that is intended to send payloads into orbit with unmatched accuracy and dependability.

The famous Tanegashima Space Centre, which is situated on a secluded island off the southern coast of Japan, is the launch site for the H3, which is equipped with strap-on solid rocket boosters and liquid thruster rockets.

The H3 is the first rocket of its sort in the world because of its creative usage of an expander bleed cycle for the first-stage engine, which sets it distinct from its predecessors.

With the help of this ground-breaking technology, the H3 can launch bigger payloads into orbit than ever before because to its increased thrust and higher fuel efficiency.

However, the H3 is more than simply a technical marvel; it is a symbol of Japan's everlasting dedication to scientific development and space exploration. The H3 takes us one step closer to discovering the mysteries of the cosmos with every successful launch, from examining far-off galaxies to comprehending the beginnings of life itself.

We will examine every facet of the H3 Launch Vehicle in an in-depth analysis, including everything from its construction and design to its launch history and beyond. We will go into the inner workings of the H3, closely analysing its mission objectives, payload capacity, and propulsion systems.

However, the H3 tale is about more than simply the technology—it's about the people who made it possible. The H3 is a monument to the strength of teamwork and human creativity, from the scientists and engineers who toil ceaselessly to refine its design to the astronauts who risk their lives to use it.

Come along with us as we set out on a cosmic adventure led by the H3 Launch Vehicle. We shall investigate the marvels of space and the limitless opportunities that exist beyond the stars together. Embark on a voyage into the unknown with us on the H3, where the sky is only the beginning.

H3 Launch Vehicle

As a medium-lift launch vehicle that can reliably and precisely send payloads into orbit, the H3 Launch Vehicle is a significant step forward for Japan's space exploration efforts.

Operation

The H3 is a highly adaptable launch vehicle that can carry payloads to Sun-synchronous orbit (SSO) and geostationary transfer orbit (GTO), among other orbits. Its main purpose is to offer a practical and economical way to launch satellites, research equipment, and other payloads into orbit.

Producer

The H3 Launch Vehicle is designed, manufactured, and operated by Mitsubishi Heavy Industries (MHI), a prominent aerospace corporation known for its proficiency in engineering and manufacturing. MHI works closely with the Japan Aerospace Exploration Agency (JAXA) to make sure the H3 satisfies the stringent requirements needed for space travel.

Origin Nation

Japan, a country with a long history of technical innovation and a strong dedication to space exploration, is the home of the H3 Launch Vehicle.

Japan is becoming a major player in the international space business because to its engineering and aerospace technological know-how; the H3 is proof of this.

Launch Cost

For sending payloads into space, the H3 offers competitive pricing with an emphasis on accessibility and affordability. With launch costs as low as US$50 million for some configurations, it is an appealing choice for a broad spectrum of clients, including academic institutions, commercial satellite operators, and governmental organisations.

Dimensions

The H3 Launch Vehicle's remarkable size guarantees that it can carry a range of payloads while still operating at peak efficiency.

Height:

The H3, commanding attention on the launch pad at a towering 63 metres (207 ft), symbolises Japan's ambition and resolve in the realm of space exploration.

Diameter:

The H3's 5.27-meter (17.3-foot) diameter allows for plenty of room for payloads to be safely stored inside its payload fairing, protecting them from harm while in orbit.

Mass:

The H3 Launch Vehicle can deliver heavy payloads into orbit because to its sturdy design, which is demonstrated by its gross mass of about 574,000 kilogrammes (1,265,000 pounds).

The H3 Launch Vehicle, a mainstay of Japan's space programme, represents the country's dedication to expanding the frontiers of scientific research and exploration.

With its cutting-edge technology, dependable operation, and economical design, the H3 is well-positioned to influence space exploration for many years to come.

Phases

The H3 Launch Vehicle consists of two independent stages, each of which is essential to launching payloads into their assigned orbits.

First Stage:

The H3's first stage produces a maximum thrust of 2,942 to 4,413 kilonewtons (661,000 to 992,000 pounds-force), depending on whether it is powered by two or three LE-9 engines. This stage provides the initial thrust required to drive the vehicle off the launch pad and into the upper atmosphere. It does this by using liquid hydrogen (LH2) and liquid oxygen (LOX) as propellants.

Second Stage:

A single LE-5B-3 engine, with a maximum thrust of 137 kilonewtons (31,000 pounds of force), powers the H3's second stage. The second stage, like the first, is propelled into space and carries payloads to their designated orbits by using LH2 and LOX as propellants.

Ability

With its remarkable payload capabilities, the H3 Launch Vehicle can carry a variety of payloads to different orbits, such as the geostationary transfer orbit (GTO) and Sun-synchronous orbit (SSO).

Payload to Sun-synchronous Orbit (SSO):
Up to 4,000 kilogrammes (8,800 pounds) of payload may be transported to Sun-synchronous orbit (SSO) by the H3. For Earth observation satellites and remote sensing missions, this orbit is ideal because it offers steady illumination for images.

Payload to Geostationary Transfer Orbit (GTO):
Depending on the particular model, the H3 has a payload capacity of between 4,800 and 7,900 kilogrammes (8,800 and 17,400 pounds) for payloads going into geostationary transfer orbit (GTO). Telecommunications satellites are frequently

placed in geostationary transfer orbit, which enables them to keep a constant altitude above the surface of the Earth.

H3 Launch Vehicle is a dependable and affordable choice for a variety of space missions because of its remarkable payload capabilities and adaptable staging configuration.

The H3 is a prime example of Japan's dedication to furthering space exploration and technology, whether it is launching satellites into orbit or funding scientific research projects.

The History

The H3 Launch Vehicle's launch history illustrates its development into a dependable and competent medium-lift launch vehicle that supports missions to many orbits with an emphasis on efficiency and accuracy.

Status:

According to the most recent information available, the H3 Launch Vehicle is still in use, carrying out its function in Japan's space programme and providing a dependable way of delivering payloads into orbit.

Launch Sites:

Tanegashima Space Centre, in Kagoshima Prefecture, Japan, is the launch site for the H3 Launch Vehicle. With its isolated position and close proximity to the Pacific Ocean, this internationally renowned spaceport offers an optimal launch environment.

Total Launches:

The H3 Launch Vehicle has carried out a total of two launches since its first voyage, signifying important turning points in Japan's space research endeavours.

Successes:

Of the two launches that were carried out, the H3 was able to complete one mission successfully, proving its dependability and efficiency in putting payloads into orbit.

Failures:

Sadly, there has only been one mission failure with the H3, underscoring the hazards and difficulties that come with space travel. Notwithstanding this obstacle, work is still being done to enhance the H3's functionality for further missions.

First Flight:

On March 7, 2023, the H3 Launch Vehicle took off, signalling a significant milestone in Japan's space programme. This test flight demonstrated the H3's capabilities and set the stage for further missions.

Last Flight:

On February 17, 2024, the H3 Launch Vehicle made its most recent flight, reaffirming its status as a dependable workhorse in Japan's space fleet. Every flight provides insightful data that improves the performance and dependability of the H3.

Type of Passengers/Cargo:

A wide variety of passengers and cargo, such as satellites, research equipment, and other payloads intended for different orbits, can be transported by the H3 Launch Vehicle. The H3 can host a wide range of payloads to satisfy the objectives of its clients, whether they are related to scientific research activities, telecommunications endeavours, or Earth observation missions.

The H3 Launch Vehicle, which symbolises Japan's dedication to exploration, innovation, and discovery, continues to be a vital component of its space programme as it pursues ambitious scientific objectives and grows its space presence. With every mission that is successful, the H3 advances humanity's collective space exploration and solidifies Japan's leadership in the international space sector.

Activators

Boosters are a feature of the H3 Launch Vehicle that increase its thrust during liftoff, improving its efficiency and dependability for launching payloads into space.

Number of Boosters:
Depending on the particular mission requirements and payload characteristics, the H3 Launch Vehicle can be built with 0, 2, or 4 boosters.

Powered by:
SRB-3 (Solid Rocket Booster-3) engines power the boosters, giving the main engines more push to help them throughout the first few stages of ascent.

Maximum Thrust:

The 2,158 kilonewtons (485,000 pounds-force) of maximum thrust produced by each SRB-3 booster considerably adds to the H3's overall thrust production during liftoff.

Specific Impulse:

The SRB-3 boosters' specific impulse, which measures around 283.6 seconds (2.781 kilometres per second), demonstrates how well they convert fuel mass into thrust.

fire Time:

The SRB-3 boosters fire for around 105 seconds, during which they give the H3 Launch Vehicle an extra push to get it off the launch pad and into the upper atmosphere.

Propeller:

Solid propellant is used to power the boosters, providing a high thrust-to-weight ratio and steady performance over the course of the burn. This kind of propellant is ideal for delivering the quick acceleration needed during the launch's early phases.

The H3 Launch Vehicle optimises its lifting capability and guarantees the safe delivery of payloads into their intended orbits by designing in boosters.

The vehicle can perform a broad range of mission profiles, from placing tiny satellites to launching heavier payloads into geostationary transfer orbit, thanks to the smart employment of boosters.

Initial Phase

The H3 Launch Vehicle's first stage is essential for getting the rocket off the launch pad and into the upper atmosphere because it provides the first push needed for ascent.

Powered by:

Depending on the exact configuration of the car, the initial stage of the H3 is powered by two or three LE-9 engines. The primary thrust needed to launch the rocket into space and get it off the ground is produced by these engines.

Maximum Thrust:

Depending on the number of engines used and the H3 Launch Vehicle model, the LE-9 engines may deliver a maximum thrust of 2,942 to 4,413 kilonewtons (661,000 to 992,000 pounds-force). The first stage's powerful thrust production allows it to break through the thick lower atmosphere and enter the almost vacuum of space.

Specific Impulse:

The LE-9 engines' efficient conversion of propellant mass into thrust is demonstrated by their specific impulse, which is around 425 seconds (4.17 kilometres per second). The initial stage's overall performance and efficacy in getting the vehicle on the intended trajectory are enhanced by this high specific impulse.

Propellant:

A mixture of liquid hydrogen (LH2) and liquid oxygen (LOX) propellants powers the H3's first stage. This mixture of propellants is ideal for use in rocket engines because it has an excellent thrust-to-weight ratio and efficient combustion properties. In addition to ensuring clean and ecologically friendly combustion, the use of LH2 and LOX propellants reduces the vehicle's operational impact on the environment.

Utilising its potent LE-9 engines and cutting-edge propellant science, the H3 Launch Vehicle's first stage supplies the thrust and propulsion required to start the rocket's ascent and send it into its target orbit. This crucial phase lays the groundwork for the mission's success by enabling the second phase to carry on with the space mission and deliver the payload to its intended location.

Phase Two

The cargo is propelled farther into space by the second stage of the H3 Launch Vehicle after the first stage completes its initial ascent, advancing the payload closer to its intended orbit.

Powered by:
One LE-5B-3 engine, which acts as the stage's main propulsion system, powers the second stage of the H3. The thrust necessary to propel the vehicle and cargo through the

upper atmosphere and into orbit is produced by this engine.

Maximum Thrust:

The H3 and its cargo may be propelled into the target orbit with the LE-5B-3 engine's maximum thrust of 137 kilonewtons (31,000 pounds-force). Even if this thrust output isn't as high as it was in the first stage, it's still enough to propel the rocket through space at the necessary speed to enter orbit.

Specific Impulse:

The LE-5B-3 engine's efficient conversion of fuel mass into thrust is demonstrated by its specific impulse, which is around 448 seconds (4.39 kilometres per second). This high specific impulse guarantees maximum efficiency and performance, enabling the second stage to run efficiently throughout the whole burn time.

Propellant:

Liquid hydrogen (LH2) and liquid oxygen (LOX) propellants are used in tandem to power the second stage of the H3, much like in the first. The high energy density and efficient combustion properties of this propellant combination allow the LE-5B-3 engine to provide consistent power and propulsion throughout the second stage burn.

The second stage of the H3 Launch Vehicle is pivotal to the effective delivery of payloads to their designated orbits and to orbital insertion by harnessing the power of the LE-5B-3 engine and cutting-edge propellant technology. The second stage, which complements the previous stage and completes the space mission, represents a significant advancement in Japan's continuous space exploration efforts.

The H3 Launch Vehicle is the result of Japan's deep experience in aerospace engineering and its dedication to innovation and technological growth in space exploration. The H3 has undergone extensive testing, optimisation, and iteration from the system's conception to its first flight, transforming it into a dependable and competent medium-lift launch vehicle.

In order to begin the development process, Mitsubishi Heavy Industries (MHI) worked with the Japan Aerospace Exploration

Agency (JAXA) to conduct comprehensive research and design. Engineers and scientists devoted countless hours to conceptualise and improve the design of the H3, making sure it would fulfil the rigorous criteria of contemporary space missions. They did this by drawing on decades of expertise in rocketry and spaceflight.

Important facets of the development of the H3 comprised:

Conception:
The H3 Launch Vehicle's anticipated capabilities, performance constraints, and mission objectives were outlined in the conception phase of the development process. The best configuration for the vehicle was determined by carefully

weighing several design possibilities and propulsion systems.

After the conceptual framework was established, engineers worked to develop and engineer the H3's first and second stages, propulsion systems, payload fairings, and guidance systems, among other parts. To verify the concept and make sure it was feasible, a lot of computer modelling and simulations were done.

Testing of Prototypes:
To evaluate the H3 Launch Vehicle's performance in simulated flight circumstances, prototypes underwent extensive testing and assessment. Before moving on with full-scale manufacturing, engineers were able to find and fix any

possible problems or design faults during this testing phase.

Production and Assembly:

Following the completion of the design, MHI started producing and assembling the H3 Launch Vehicle at sites spread throughout Japan. To guarantee the dependability and integrity of every component, quality assurance procedures and precision engineering were put in place.

Integration and Testing:

After each component was finished, it was assembled into the H3 Launch Vehicle's final assembly. Static firing tests and subsystem checks were among the integration tests that were carried out to confirm the vehicle's performance and operation once it was fully completed.

Launch missions:

At the Tanegashima Space Centre, the H3 Launch Vehicle participated in a number of launch missions after integrating and testing successfully. To guarantee a smooth and successful launch operation, these programmes included pre-launch preparations, fuelling procedures, and last countdown drills.

Operational Deployment:

The H3 Launch Vehicle entered operational deployment as a dependable and affordable way to launch payloads into orbit following the conclusion of launch campaigns and a successful first flight. The success of the H3 missions is ensured by ongoing maintenance and monitoring.

The success of the H3 Launch Vehicle was largely due to the cooperation of MHI, JAXA, and other industrial partners during the development phase. The H3 is a major advancement in Japan's efforts to explore space and make scientific discoveries by utilising the country's competence in aeronautical engineering and technology.

The H3 will live on as a tribute to human ingenuity and the spirit of exploration as Japan pushes the bounds of what is feasible in space.

A variety of variations of the H3 Launch Vehicle are available to fulfil the various requirements of space missions, ranging from satellite deployment to scientific research and exploration missions.

Every variation is designed to efficiently and precisely transport payloads to designated orbits, demonstrating the H3 platform's adaptability and versatility. Here are a few noteworthy variations:

H3-30S/L:

Missions needing to carry payloads to Sun-synchronous orbit (SSO) are intended for this type. The H3-30S/L model is perfect for Earth observation satellites, remote sensing missions, and environmental monitoring programmes since it can carry payloads up to 4,000 kilogrammes (8,800 pounds).

H3-24S/L:

This version of the aircraft is designed to fly to geostationary transfer orbit (GTO) and has a cargo capacity of 4,800 to 7,900 kg (8,800 to 17,400 pounds). For commercial satellite installations that need accurate geostationary orbital placement, television platforms, and telecommunications satellites are good candidates for this type.

H3-24:

Specifically engineered for lunar missions, the H3-24 version provides increased payload capacities for payloads headed for lunar transfer orbit (TLI). The H3-24 model makes a wide range of lunar exploration and research efforts possible with its capacity to carry more than 6,000 kilogrammes (13,000 pounds) of cargo to TLI and 8,800 kilogrammes (19,400 pounds) to geostationary transfer orbit (GTO).

Custom Configurations:
The H3 Launch Vehicle may be tailored to fulfil certain mission requirements and payload specifications, in addition to the regular variations.

Because of its adaptability, customised solutions may be developed to meet the needs of various payloads, launch trajectories, and mission goals, guaranteeing successful mission execution and optimal performance.

With the integration of cutting-edge technologies, propulsion systems, and payload accommodations, every H3 Launch Vehicle variation expands upon the fundamental capabilities of the platform to provide unmatched performance and dependability.

The H3 is a flexible and affordable option for a variety of space missions, supporting Japan's aspirational objectives in space exploration and scientific research thanks to its customisable configurations and versatile architecture.

Conclusion

Japan's creativity, inventiveness, and dedication to space exploration are demonstrated by the H3 Launch Vehicle. The H3, with its state-of-the-art technology, sturdy construction, and adaptable functions, is a promise for future exploration efforts and a major turning point in Japan's space programme.

The H3 has shown to be incredibly reliable and performant, carrying payloads to different orbits with efficiency and accuracy from its first flight. Its powerful propulsion systems, modular architecture, and configurable configurations enable it to be a versatile and adaptable platform that can serve a broad range of space missions, from lunar exploration endeavours to satellite deployments.

The future of humanity's space travel will be significantly shaped by Japan's H3 Launch Vehicle as it continues to push the frontiers of space research.

The H3 takes us one step closer to solving the universe's riddles and deepening our knowledge of it with every successful launch.

The H3 will keep setting the standard in the upcoming years as new technologies appear and uncharted territories call for exploration, motivating upcoming generations of engineers, scientists, and explorers to aim high.

The H3 Launch Vehicle, a representation of humanity's pursuit of knowledge and exploration, epitomises the finest aspects of Japan's space programme with its steadfast commitment to quality and pioneering spirit.